Science Inquiry

Living Things

by Joe Baron

Science Inquiry

Science in a Snap!

Explore Activity

Investigate the Needs of Animals 6

▶ **Question:** What do pill bugs need to stay alive?

Think Like a Scientist **Math in Science** 10

Tables

Directed Inquiry

Investigate Plants and Water 14

▶ **Question:** What happens if radish plants
do not get water?

Science in a Snap!

Things All Around

Make a table. Write **Living** and **Nonliving.** Look around the room and out the window. Draw 4 living things you see. Then draw 4 nonliving things. Show your table. Tell why you chose the things you drew.

Plant Parts Match

Draw a plant. Show the roots, stem, leaves, and flower. Write words for plant parts on cards. Match the words and plant parts. How do the roots, stem, and leaves help the plant?

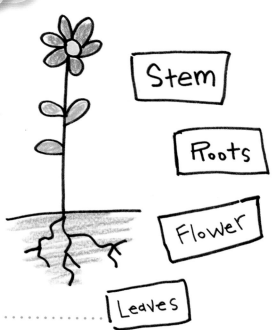

Do You Need It?

Write **Yes** and **No** on 2 sheets of paper. Hold up **Yes** each time your teacher names something that people need to live. Hold up **No** for something you do not need to stay alive. Tell how people get one thing they need to live.

Investigate the Needs of Animals

Question What do pill bugs need to stay alive?

Science Process Vocabulary

observe verb

When you **observe** something, you use your senses to learn about it.

infer verb

When you **infer,** you use what you know and what you observe to explain something.

I infer that the pill bugs are eating the leaf.

6

Materials

pan with soil

apple pieces

spoon and water

safety goggles

lettuce leaves

paper towel

pill bugs

What to Do

1 Put on safety goggles. Put some leaves and apple pieces on top of the soil.

2 Make a small ball from the paper towel. Put it in the pan. Place 1 spoonful of water on the paper towel.

3 Put the pill bugs in the pan.

4 **Observe** the pill bugs every day for 3 days. What do you see them do? Look for clues about how pill bugs get food and water. What can you **infer** about the pill bugs from what you observe? Write in your science notebook.

Record

Write or draw in your science notebook. Use a table like this one.

Day	I Observe	I Infer
1		
2		
3		

Share Results

1. Tell what you did.

> I observed _____.

2. **Infer** where the pill bugs get food and water.

> Pill bugs get food from _____.
> They get water from _____.

Math in Science

Tables

You can use a table to organize and compare data. Tim wanted to know how many things he saw in a park were living and how many were nonliving. Tim started to make a list.

Tim's list did not tell how many rocks and trees he saw. So he made a mark each time he saw one.

rocks

| |

trees

| | |

Then Tim decided that a table might work better. A table can have these parts:

- The **title** tells you what the table is about.

- **Labels** tell you how the data are sorted.

- The **rows** and **columns** tell you where to put the data.

- The empty boxes are for your data.

Title

	Label A	Label B	Label C
row 1 ➡			
row 2 ➡			
row 3 ➡			

⬆ column 1   ⬆ column 2   ⬆ column 3

Tim used a table to sort what he saw. In column 1, Tim recorded the things he saw.

Tim wrote a mark in column 2 if what he saw was a living thing. He put a mark in column 3 if it was nonliving.

Things in the Park		
What I Saw	Living	Nonliving
Rock		I I I
Tree	I I	

▶ **What Did You Find Out?**

1. How many living things did Tim see? How many nonliving things did he see?

2. How does the table help make information easier to understand?

 Make and Use a Table

1. Look for living and nonliving things
    in the picture. Use a table to record
    what you see.

    - Write a **title.**

    - Write a **label** for each column.

    - Write the names of things in
      column 1.

    - Write a mark to show whether
      each thing is living or nonliving.

2. Share your table with others and
    ask questions about it.

Investigate Plants and Water

 **Question** What happens if radish plants do not get water?

Science Process Vocabulary

observe verb

You can **observe** an object by using senses such as smell and sight.

predict verb

When you **predict,** you say what you think will happen.

I predict that some plants will grow faster than others.

Materials

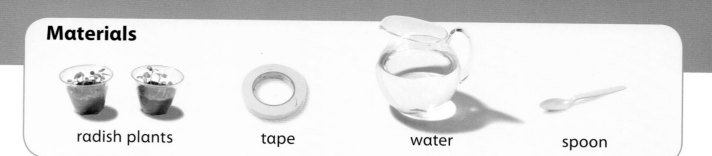

radish plants   tape   water   spoon

What to Do

1 Write **Water** on one cup.
Write **No water** on the other cup.

2 Put 2 spoonfuls of water on the plants in the cup labeled **Water**. Do not put water on the plants in the other cup.

3 **Observe** the plants. Record your observations in your science notebook.

4 Place the plants in a sunny place.

5 Put 2 spoonfuls of water in the cup labeled **Water** every day. Do not put water in the other cup. **Predict** in which cup the plants will grow better. Write in your science notebook.

6 Observe your plants every day for 5 days. Record your observations.

Record

Write or draw in your science notebook. Use a table like this one.

Day	Water	No Water
1	 water	No water
2		

Explain and Conclude

1. Do your results support your **prediction?** Explain.
2. Tell what you conclude about water and plants.

Think of Another Question

What else would you like to find out about plants and water?

Investigate Plants and Light

Question   What happens if radish plants do not get light?

Science Process Vocabulary

share verb

When you **share** results, you tell or show what you have learned.

infer verb

When you **infer,** you use what you know and what you observe to explain something.

I infer that this plant did not get its basic needs met.

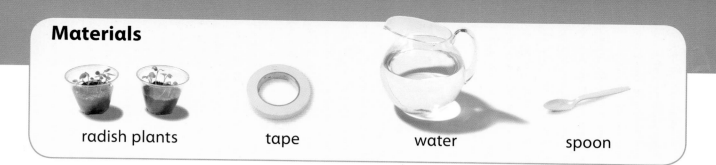

radish plants   tape   water   spoon

What to Do

1 Write **Sunlight** on one cup. Write **No sunlight** on the other cup.

2 **Observe** the plants in both cups. Record your observations in your science notebook.

What to Do, continued

3 Place the **Sunlight** cup in a sunny place. Place the **No sunlight** cup in a dark place.

4 **Predict** in which cup the plants will grow better. Write in your science notebook.

5 Give each cup of plants 2 spoonfuls of water every day. Observe the plants every day. Record what you observe.

Record

Write or draw in your science notebook. Use a table like this one.

Day	Sunlight	No Sunlight
1		
2		

Explain and Conclude

1. **Infer** why the plants in the 2 cups grew differently.
2. **Share** your results with the class. Tell what you conclude about light and plants.

Think of Another Question

What else would you like to find out about plants and light?

Investigate Plants and Space

Question What happens if bean plants or sunflower plants are crowded?

Science Process Vocabulary

predict verb

When you **predict,** you use what you know to say what will happen.

> I predict that the seeds will grow into plants.

fair test

An investigation is a **fair test** if you change only one thing and keep everything else the same.

Materials

12 bean seeds

12 sunflower seeds

water

2 cups with soil

tape

safety goggles

spoon

Do a Fair Test

Write your plan in your science notebook.

Make a Prediction

In this investigation, you will give some plants a lot of space. You will give other plants only a little space. **Predict** which plants will grow taller.

Plan a Fair Test

What one thing will you change?
What will you observe or measure?
What will you keep the same?

What to Do

1 Put on your safety goggles.
Label one cup **Crowded.**
Label the other cup **Not crowded.**

2 Choose a kind of seed to test. Put 10 of the seeds in the **Crowded** cup. Put 2 of the seeds in the **Not crowded** cup.

3 Add the same amount of soil to each cup. Cover the seeds.

4 Put 4 spoonfuls of water in each cup.

5 Place both cups in a sunny place. **Observe** the plants every day. Write what you observe in your science notebook.

6 **Compare** your results with the results of other groups. Look for **patterns.**

Record

Write or draw in your science notebook. Use a table like this one.

Kind of Seed _____

Day	Crowded	Not Crowded
1		
2		

Explain and Conclude

1. Do the results support your **prediction?** Explain.
2. What **patterns** did you find when you **compared** data from other groups?
3. Tell what you **conclude** about plants and the space they need to grow.

Think of Another Question

What else would you like to find out about what plants need to grow? What could you do to answer this new question?

Investigate Basic Needs of Humans

Question What happens to the number of breaths you take as you run longer?

Science Process Vocabulary

data noun

You collect **data** when you gather information in an investigation.

	Number of Breaths
1 min	15
2 min	20
3 min	22

patterns noun

Look for **patterns** in your data, such as numbers, that repeat or change in a regular way.

Materials

stopwatch

What to Do

1 Sit quietly in a chair.
Have a partner use the
stopwatch to time you
for 1 minute.

2 Then have a partner count how many
breaths you take in 30 seconds.
Record the **data**
in your science
notebook.

3 Run fast in place. Have a partner tell when you have run for 30 seconds. Then do step 4 right away.

4 Count how many breaths you take in 30 seconds. Record the data.

5 Rest for 1 minute. Then do steps 1, 2, 3, and 4 again. This time run in place for a longer time. Decide whether you will run for 1 minute or 2 minutes.

Record

Write in your science notebook.
Use a table like this one.

	Before Running	Running 30 seconds	Running _____
Number of Breaths			

My Science Notebook

Make a graph.

Running Time and Breaths

Number of Breaths	Before Running	Running 30 seconds	Running _____
12			
11			
10			
9			
8			
7			
6			
5			

Explain and Conclude

1. Look for **patterns** in your data. How did the number of breaths change as you ran longer?

2. When you breathe, your body gets the air it needs. **Infer** why the number of breaths changed as you ran longer.

Think of Another Question

What else would you like to find out about the basic needs of humans?

Do Your Own Investigation

Question Choose a question, or make up one of your own to do your investigation.

- What happens if seeds do not get sunlight?
- What happens if plants get plant food?
- What happens to the number of breaths you take as you run faster?

Science Process Vocabulary

fair test

An investigation is a **fair test** if you change only one thing and keep everything else the same.

Open Inquiry Checklist

Here is a checklist you can use when you investigate.

- ☐ Choose a **question** or make up one of your own.

- ☐ Gather the materials you will use.

- ☐ Tell what you **predict.**

- ☐ Plan a **fair test.**

- ☐ Make a **plan** for your investigation.

- ☐ Carry out your **plan.**

- ☐ Collect and record **data.** Look for **patterns** in your data.

- ☐ Explain and **share** your results.

- ☐ Tell what you **conclude.**

- ☐ Think of another question.

How Scientists Work

Investigating Again and Again

When scientists do investigations, they carefully record what they learn. They do the tests again and again. They do the tests the same way each time. That way they can be sure of the results.

▲ This scientist learns about Everglades plants by observing and testing.

Plants need nutrients. In the Florida Everglades, the amount of nutrients has changed. Now cattails grow in some places where sawgrass used to grow.

◄ Cattails grow in places with a lot of nutrients.

Sawgrass grows ► in places with little nutrients.

continued

Scientists investigate changes in nutrients in the Everglades. They observe the plants growing there. They learn how the plants have changed. Scientists also do fair tests to learn about changes in nutrients in the Everglades.

◄ **This scientist learns about Everglades plants by observing the plants where they grow.**

► **What Did You Find Out?**

1. How do scientists learn about nutrients and plant growth in the Florida Everglades?

2. Why do scientists do fair tests again and again?

 Repeat an Investigation

1. With a team, investigate to find out which sprouts first—sunflower seeds or bean seeds. Do the investigation the same way three times.

   • Explain what you did in your investigation.

   • Compare your results with the investigations of the other teams.

2. Explain why you would want to get the same results each time.

featured photos

Cover: giraffes standing together, Africa

Title page: plant bug looks like an ant through mimicry

page 9: pill bugs, armadillidium vulgare

inside back cover: horses running on grass near stream and hills, North America

ACKNOWLEDGMENTS
Grateful acknowledgment is given to the authors, artists, photographers, museums, publishers, and agents for permission to reprint copyrighted material. Every effort has been made to secure the appropriate permission. If any omissions have been made or if corrections are required, please contact the Publisher.

PHOTOGRAPHIC CREDITS:
set-up photography: Andrew Northrup; stock photography: Cover (bg) Digital Stock/Corbis; Title (bg) Robert Sisson/National Geographic Image Collection; 6 (t) Wayne Bennett/Corbis, (b) James H. Robinson/Photo Researchers, Inc.; 9 Nature's Images/Photo Researchers, Inc.; 10-11 (bg) 22DigiTal/Alamy Images; 13 blickwinkel/Schmidbauer/Alamy Images; 14 (t) Digital Stock/Corbis, (b) B. Bird/zefa/Corbis; 17 David Cook / blueshiftstudios/Alamy Images; 18 (t) BrandX/Jupiterimages, (b) Nigel Cattlin/Alamy Images; 21 silver-john/Shutterstock; 22 (t) Ed Young/AgStock Images/Corbis; 22 (b) Randy Faris/2007/Corbis; 25 Ariel Skelley/Corbis; 26 (t) Bloomimage/Corbis, (b) LWA-Dann Tardif/Corbis; 27 (tr) Erik Isakson/Corbis; 30 (t) Maddie Thornhill/Alamy Images, (b) Randy Faris/Corbis; 31 Martin Gallagher/Corbis; 32 David R. Frazier/Photo Researchers, Inc.; 32-33 (bg), 33 (t) Tim Fitzharris/Minden Pictures/National Geographic Image Collection; 33 (b) David Muench/Corbis; 34 Robert Sisson/National Geographic Image Collection; 35 (l) David Doubilet/National Geographic Image Collection, (r) James D Watt/Stephen Frink Collection/Alamy Images; Inside Back Cover (bg) Digital Stock/Corbis.

ILLUSTRATIONS: Amy Loeffler

PUBLISHED BY NATIONAL GEOGRAPHIC SCHOOL PUBLISHING
& HAMPTON-BROWN
Sheron Long, Chairman
Samuel Gesumaria, Vice-Chairman
Alison Wagner, President and CEO
Susan Schaffrath, Executive Vice President, Product Development

Editorial: Fawn Bailey, Joseph Baron, Carl Benoit, Jennifer Cocson, Francis Downey, Richard Easby, Mary Clare Goller, Chris Jaeggi, Carol Kotlarczyk, Kathleen Lally, Henry Layne, Allison Lim, Taunya Nesin, Paul Osborn; Chris Siegel, Sara Turner, Lara Winegar, Barbara Wood.

Art, Design, and Production: Andrea Cockrum, Kim Cockrum, Adriana Cordero, Darius Detwiler, Alicia DiPiero, David Dumo, Jean Elam, Jeri Gibson, Shanin Glenn, Raymond Godfrey, Raymond Hoffmeyer, Rick Holcomb, Cynthia Lee, Anna Matras, Gordon McAlpin, Melina Meltzer, Rick Morrison, Christiana Overman, Andrea Pastrano-Tamez, Leonard Pierce, Cathy Revers, Stephanie Rice, Christopher Roy, Janet Sandbach, Susan Scheuer, Margaret Sidlosky, Jonni Stains, Shane Tackett, Andrea Thompson, Andrea Troxel, Ana Vela, Teri Wilson, Chaos Factory, Feldman and Associates, Inc., Tighe Publishing Services, Inc.

THE NATIONAL GEOGRAPHIC SOCIETY
John M. Fahey, Jr., President & Chief Executive Officer
Gilbert M. Grosvenor, Chairman of the Board

MANUFACTURING AND QUALITY MANAGEMENT,
The National Geographic Society
Christoper A. Liedel, Chief Financial Officer
George Bounelis, Vice President

National Geographic School Publishing
Hampton-Brown
P.O. Box 223220
Carmel, California 93922
www.NGSP.com

Printed in the USA.

ISBN: 978-0-7362-6224-8

10 11 12 13 14 15 16 17

10 9 8 7 6 5 4 3 2 1